CINCINNATI ZOO AND BOTANICAL GARDEN

DAVID A. OEHLER

The Rosen Publishing Group's
PowerKids Press™
New York

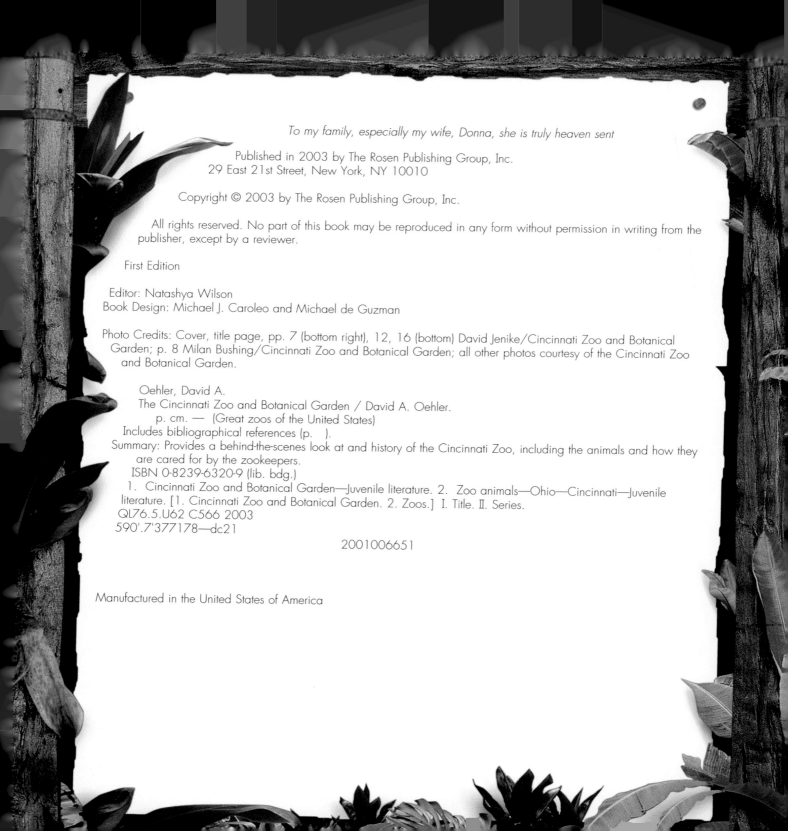

To my family, especially my wife, Donna, she is truly heaven sent

Published in 2003 by The Rosen Publishing Group, Inc.
29 East 21st Street, New York, NY 10010

First Edition

Editor: Natashya Wilson
Book Design: Michael J. Caroleo and Michael de Guzman

Photo Credits: Cover, title page, pp. 7 (bottom right), 12, 16 (bottom) David Jenike/Cincinnati Zoo and Botanical Garden; p. 8 Milan Bushing/Cincinnati Zoo and Botanical Garden; all other photos courtesy of the Cincinnati Zoo and Botanical Garden.

Oehler, David A.
The Cincinnati Zoo and Botanical Garden / David A. Oehler.
    p. cm. — (Great zoos of the United States)
Includes bibliographical references (p.   ).
Summary: Provides a behind-the-scenes look at and history of the Cincinnati Zoo, including the animals and how they are cared for by the zookeepers.
ISBN 0-8239-6320-9 (lib. bdg.)
1. Cincinnati Zoo and Botanical Garden—Juvenile literature. 2. Zoo animals—Ohio—Cincinnati—Juvenile literature. [1. Cincinnati Zoo and Botanical Garden. 2. Zoos.] I. Title. II. Series.
QL76.5.U62 C566 2003
590'.7'377178—dc21

                          2001006651

Manufactured in the United States of America

Contents

In 1898, many families rode cable cars to the Zoo. To get into the Zoo, they paid 25 cents for adults and 10 cents for kids.

The world's last passenger pigeon, Martha, lived at the Zoo. She died in 1914.

This is Andrew Erkenbrecher, the founder of the Cincinnati Zoo.

The former Monkey House is the oldest zoo building in America. Today it is the Reptile House.

This bird's-eye view of the Zoo grounds was taken in the 1870s, when the Zoo was being built.

In 1875, a man named Andrew Erkenbrecher lived in Cincinnati, Ohio. For more than two years, he had worked to fulfill his dream of building the world's finest zoo. On September 18, 1875, the Cincinnati Zoo opened. More than 1,000 people visited the Zoo that day. It was the second zoo to open in the United States. The Philadelphia Zoo was the first. Two of the Cincinnati Zoo's first displays were the Monkey House and the Bird **Aviaries**. These buildings are still in use. One of the aviaries is now a **monument** to Martha, the world's last passenger pigeon. The Monkey House is now the Reptile House. It is the oldest zoo building in America. In 1974, the Zoo was named a national historic landmark.

5

The Cincinnati Zoo is a place to have fun and to learn. More than one million people visit the Zoo each year. Its 75 acres (30 ha) are located in central Cincinnati. More than 700 **species** of animals and 3,000 species of plants live in the Zoo. The Children's Zoo is located in the middle of the Zoo. Elephants live to the south, gorillas to the west, rhinos to the north, and birds to the east. There are many more animals, too! It takes more than 350 employees and 1,100 **volunteers** to run the Zoo. Shows teach visitors about animal behavior and wildlife **conservation**. Handlers bring animals to schools to teach students about wildlife. The Cincinnati Zoo Conservation Fund works to help animals around the world!

Thousands of animals and plants live at the Cincinnati Zoo! Take a look at the map below to see all the things there are to visit.

VINE ST. (to I-75)

CREW
Carl H. Lindner, Jr. Family Center for Conservation and Research of Endangered Wildlife

CREW Endangered Species Garden

CREW Endangered Species Garden

Gorilla World

Passenger Pigeon Memorial

Nocturnal House

Peacock Pavilion

Lodge

Whiting Grove Picnic Area

Peacock Parking Lot

Insect World

Komodo Dragon

Butterfly Garden

Wildlife Canyon
Home of Emi, the Sumatran rhino

Dinosaur Garden
(between Reptile House and Monkey Island)

Cat House

Lemur Exhibit

GROVE SHOP

Manatee Springs
presented by the Otto M. Budig Family Foundation

Parking Lot

Metro Bus Stop

Eagle Eyrie

Reptile House

Monkey Island

Cat Grottoes

Pollinator Garden

Amphitheater

Zoo Gallery

Rhino Reserve

Big Cat Canyon

Red Panda

Veldt

Pedestrian Entrance and Exit

ELEPHANT EMPORIUM

Blakley's Barn

African Flamingo

Cheetahs

Otter Creek

POLAR BEAR GIFT SHOP

Kroger Lords of the Arctic

African Bldg.

Jungle Trails

Giraffe

Offices

Wetland Trails

Gibbon Islands

Joseph H. Spaulding, Jr. Children's Zoo

Australian Walkabout

Seals

Bears

Asian Elephants

Okapi

Swan Lake

KIDS SHOP

Marge Schott-Unnewehr Vanishing Giants

Elephant Performance Yard

Train Station and Tram Stop

Garden of Peace

Upper Penguin Parking Lot

Oriental Garden

ZOO SHOP

Safari Restaurant

Walk-through Flight Cage

Wings Of The World

Asian Bldg.

Auto Exit

Frisch's Discovery Center

Education Building

Botanical Center

Main Entrance

THORN TREE SHOP

Penguin Parking Lot

Frisch's Animal Recreation Center

Marketing, Human Resources & Business Offices

Auto Entrance

Exit

Pedestrian Entrance and Exit

DURY AVE.

Camel Parking Lot

ERKENBRECHER AVE.

FOREST AVE.

katherine kremer design, inc.

This beautiful macaw, a type of parrot, is just one of the many birds that visitors can see in the Wings of the World display.

Walkingstick insects are herbivores, which means they eat only plants. Their sticklike bodies help them to hide in bushes to stay safe.

Bugs are everywhere! Most bugs are actually called insects. They fly, crawl, jump, and climb. The World of Insects is in the northwest corner of the Cincinnati Zoo. It houses walkingsticks, diving beetles, ants, and many other types of insects. There is even a butterfly aviary where more than 200 butterflies fly around while visitors walk through the display. A special camera in the World of Insects shows a close-up view of leaf-cutter ants. These ants are called the farmers of the insect world. They cut pieces from leaves and take them under ground, where **fungus** grows on the pieces. They also take out old leaves and make sure there is enough fungus for all the ants to eat.

There is water almost everywhere at the Cincinnati Zoo. Northeast of the World of Insects is Manatee Springs. The Zoo's Florida manatees, Stoneman and Douglas, swim in a pool filled with 120,000 gallons (454,249 l) of water. In the wild, Florida manatees are having a tough time. They often get injured or killed by fishing nets and boats. The Zoo has joined with the Sirenia Project to track the wild manatees. The project's goal is to make the manatees' main living areas safe from boats.

East of the manatees, polar bears swim in their own pool. The Zoo's penguins also have plenty of water. There are waterfalls and pools where animals can play. In the Children's Zoo, kids can play in water, too!

The Zoo has a large grocery list! Among the items bought every year are 96,000 apples and 3,380 bales of alfalfa hay.

Special underwater views let visitors get nose to nose with the Zoo's giant manatees. Stoneman (below) and Douglas each eat 80 pounds (36 kg) of romaine lettuce every day!

This bonobo has a safe home in the Jungle Trails display. The Zoo is working to buy real rain forest land to protect the land.

Jungle Trails, Gorilla World, and Marge Schott-Unnewehr Vanishing Giants are the Cincinnati Zoo's **immersion displays**. These displays make visitors feel as if they are walking through **tropical rain forests**. The plants and the animals in them come from Africa and Asia. In Jungle Trails, visitors can see bonobos and lemurs. Gorilla World houses lowland gorillas and monkeys. Asian elephants live in the Vanishing Giants display. Asian elephants are **endangered**, because people **poach** them and cut down the rain forests where they live. Rain forests cover less than one-fourth of Earth's surface, but more than half of all plant and animal species on Earth live in them. They can get 200 inches (508 cm) of rain per year!

The Cincinnati Zoo and Botanical Garden was one of the country's first public gardens. "Botanical Garden" was added to the Cincinnati Zoo's name in 1987. Many colorful gardens decorate the Zoo. Some of the plants are types that were around when dinosaurs were alive! These plants can be found in the Dinosaur Garden, near the Reptile House. This garden also has **petrified** wood and coal, both of which are formed by really old plants.

One of the many important tasks done at the Zoo's Center for Conservation and **Research** of Endangered Wildlife is the freezing of seeds and other parts of endangered plants to save them for the future. After the plant parts are carefully thawed, they can be replanted.

Colorful flowers bloom outside the Reptile House,
the oldest zoo building in America.

15

DID YOU KNOW?

Humans are pregnant for 9 months. Rhinos are pregnant for about 15 months. Elephants are pregnant for almost 24 months!

Gorillas have babies about once every four years. Gorilla babies need a lot of care.

This Indochinese tiger was born and raised at the Zoo. Between 1,000 and 2,000 Indochinese tigers live in the wild today.

Emi's calf, Andalas, is the first Sumatran rhinoceros to be bred and born in captivity in 112 years.

16

One very special animal lives in Wildlife Canyon. Her name is Emi, and she is a **Sumatran rhinoceros**. There are only 300 Sumatran rhinos left in the world. Emi came to the Zoo in 1991, after the forest where she lived in Southeast Asia was cut down. On September 13, 2001, Emi gave birth to a calf. This is the first time since 1889 that a Sumatran rhino has been bred and born in a zoo. The birth is a big step toward saving these rhinos from becoming **extinct**. **Captive breeding** is one way the Zoo helps to save endangered wildlife. Dozens of gorillas and Indochinese tigers have been born at the Zoo. Some species, such as a bird called the Guam rail, have had their offspring put back into the wild.

WINGS AND THINGS

The Wings of the World **exhibit** houses animals with wings. There are bats, butterflies, and birds. Butterflies are insects. Bats have fur, so they are **mammals**. Birds are different from all other animals because they have feathers. Feathers keep birds warm, even in cold water. Wings of the World is filled with birds from many places. There are rhinoceros hornbills from the rain forests of Asia. They are huge birds with long, orange beaks. There are loud, thick-billed parrots from the mountains of Arizona and Mexico. Cincinnati is the only zoo to keep least auklets and whiskered auklets, birds that live on land in the North Pacific area. The Zoo also works with auklets in the wild, to keep track of their numbers and their health.

A parrot (left), a fruit bat (top right) and a butterfly (bottom) are just a few of the animals that live in Wings of the World.

19

The Zoo hopes to help save cheetahs through captive breeding. Cheetahs are the world's fastest land animals.

MARTHA'S MESSAGE

When the Europeans first came to America, there were billions of passenger pigeons. Overhunting and the cutting down of forests made them extinct within 50 years. Martha, the last passenger pigeon, died at the Cincinnati Zoo in 1914. Her death shows us that saving animals' natural homes and limiting hunting is necessary for their survival. The Cincinnati Zoo works to save wildlife and wild places. The cheetah is one endangered species that the Zoo is fighting to help. Cheetahs have been hunted for hundreds of years. Their homes in Asia and Africa are being destroyed. The Zoo is working with the Cheetah Conservation Fund in Africa to save the land where some of the last cheetahs live.

HOPE FOR THE FUTURE

The Cincinnati Zoo and Botanical Garden plans to build other exciting and educational exhibits. Medicine Trails will be one of the next exhibits built at the Zoo. It will display plants that can be made into medicines and will teach visitors how important plants are to life on Earth. Zoo scientists will study the plants and will teach visitors about the work being done in the exhibit. Medicine Trails will also explain the importance of the wildlife in rain forests.

Zoos are an important part of conservation work. In the future, instead of trying to save one animal or plant, zoos will work to save communities of plants and animals, called **ecosystems**. The Cincinnati Zoo will continue to be a place to learn about all living things.

Glossary

aviaries (AY-vee-ehr-eez) Places where flying animals are kept.

captive breeding (KAP-tihv BREED-ing) Bringing animals together to have babies in a zoo or an aquarium instead of in the wild.

conservation (kon-sur-VAY-shun) Protecting natural areas where plants and animals live.

ecosystems (EE-koh-sis-temz) The way plants and animals live together in nature and form basic units of the environment.

endangered (en-DAYN-jerd) In danger of dying out.

exhibit (ig-ZIH-buht) A display designed for people to come and see.

extinct (ek-STINKT) No longer existing.

fungus (FUN-gis) A mushroom, mold, mildew, or related living thing.

immersion displays (ih-MER-zhun dis-PLAYZ) Displays where visitors feel as if they are walking through animals' natural homes.

mammals (MA-mulz) Warm-blooded animals that have a backbone and hair, breathe air, and feed milk to their young.

monument (MAHN-yoo-mint) A place that honors a person or an event.

petrified (PEH-truh-fyd) Turned into stone.

poach (POHCH) To hunt animals when it is against the law.

research (rih-SURCH) To study something.

species (SPEE-sheez) A single kind of plant or animal.

Sumatran rhinoceros (soo-MA-trun ry-NAH-ser-uhs) The smallest and hairiest of the rhinos. They live in Sumatra, Malaysia, and Borneo.

tropical rain forests (TRAH-pih-kul RAYN FOR-ests) Areas near Earth's equator that are always warm and get large amounts of rain.

volunteers (vah-luhn-TEERZ) People who offer to work for no pay.

Index

Web Sites

To learn more about the Cincinnati Zoo and Botanical Garden, check out this Web site:

www.cincinnatizoo.org